Still Healing

A DOCTOR'S STORY OF ABORTION

by Marissa Ogle, M.D.

For Hope, who went to Heaven
December 18, 1982.

Acknowledgements

I am so very grateful...

To Kathleen, thank you for your commitment to life. In so doing you have given us the gift of a wonderful husband and amazing father. God has used your choice to begin a new family legacy.

To my family, each of whom has been an essential part of encouraging me to share the vulnerabilities and victories of my testimony, always believing that I had an important message. Tim, for never letting me quit, even when I was certain that I couldn't make it work. Your words have always been timely and spoken in love. Alex, you are a representative of God's redemption in my life, and faithfully walk out mercy, grace and truth in a way that encourages me immensely. Mom, for covering me in prayer like only a mother can, and loving me deeply through every season and endeavor. April, for inspiring me to be creative and faithful in sharing God's work in my life. Jenny, for journeying with me through valleys and over mountaintops, always with the love and hope that life in Christ brings.

To the friends who've been on this journey with me: Anne Severance and Nancy Campbell, who helped me to believe that my unique perspective as a physician needed a platform; Chris and Renay Holloway, for encouraging me to share my testimony for God's glory; Teresa Bennett, Karyn Guelcher and Annette Delk, who have lovingly read the many manuscripts along the way. Thank you all for the wellspring of encouragement and words of life. They have been oxygen to my soul.

To those who contributed voluntarily to the content of this book, it would not be a reality without you. Dr. Joseph DeCook, for reviewing both medical and spiritual content for truth and accuracy, bringing wisdom that comes from a lifetime of caring for women and their unborn children. Teasi Cannon, your words of wisdom and valuable perspective as an author were essential and timely. Kathy Troccoli, for believing in this from the very first oh-so-rough draft, and honoring that by writing the humorous yet beautifully moving forward. The authors of the anonymous letters, thank you for your willingness to go to that tender and vulnerable place in your lives for the glory of God. Breanna Butterworth, for working so diligently on the cover, overall layout, and submission process. Your passion and creativity re-ignited the fire needed to make this vision a reality.

To God, my redeemer, my healer, my peace....but most of all, my DADDY. Thank you for loving your little girl so much that You desire to conform her into the image of Christ and use her as a vessel. That You don't redeem by patching the holes, but by making us WHOLE. That You have a purpose for my life, one that begins with, and will forever remain, HOPE.

Forward

My gyno appointment. Ugh...by far my least favorite medical appointment next to the dentist. You have to deal with all kinds of body shame as you are told to put the "open in the front" towelette around your body. You lie on a medical "lazyboy" and try not to freeze to death as you anticipate putting your private parts on display for the entire world to see. Ok... well maybe just your gynecologist...and the nurse. But still, it feels like the whole world. Anyway—because cancer runs in my family, (both parents died at early ages due to this vicious disease) I approach these appointments like a skittish little girl going into a haunted house. What will I hear? What will I see?

Dr. Marissa Ogle was highly recommended to me and being religious in my check-ups I went obediently, albeit with ball and chain. She walked in my room like an old friend. Pretty, confident and caring. I felt immediately at ease. I proceeded to put my feet in those dreaded stirrups—as I heard those familiar words—"please scoot down". Conversation ensued—mostly medically informative—with a little friendly banter. I left thinking that I had just met a very fine doctor as well as a very unique woman.

The next couple of times I visited Dr. Ogle I lay in the "scooted down" position for thirty minutes—we talked and shared as if it were a casual get together at Starbucks. I was well aware of my "vulnerability" but was comfortable to just enter into conversation; I knew she would get to examine me before long. After several visits like that, I decided I'd ask her to go to dinner and that it would be much more enjoyable for me to be sitting in a chair.

Well...needless to say, Marissa has become a dear friend. She is by far one of the most remarkable women I have ever known. Her intelligence has caused her to have a unique inquisitiveness—possessing knowledge without arrogance. Her love for God has given her a love for people. Her own choices and immense suffering have yielded grace and mercy—compassion and empathy. What a lovely soul.

Dr. Ogle has an unceasing zeal to see women free of guilt and shame. See their broken-hearts become whole. See their half-lived lives become full and thriving.

This book is a gift from heaven. It is a reminder of the amazing love and forgiving-ness of God. So, go ahead. Don't be afraid. Take the journey with Marissa. We are all in process and still...healing.

Kathy Troccoli

Award-winning Singer and Songwriter

Floating.
Gently. Quietly. Peacefully.
Knowing we are one, hearing the heartbeat of my mom.

Whirring.
Noise. Pulling. Tugging! PAIN, OH, Pain!!!! Cold, bright light.
Darkness. Quiet. Despair. Where did she go?

Floating.
Up and up! Brilliant light. Joy, overwhelming joy! Peace...
Now again as one...
Hearing the heartbeat of the Son.

Introduction

As a physician, I have the opportunity to listen to women one-on-one, face-to-face, as they share freely with me. I've had the opportunity to meet many women whose lives have been forever changed by abortion, promiscuity, sexually transmitted diseases and unplanned pregnancy. So many who have bought into the massive lies of our culture—including that our very worth can be quantified by another human being. That lie and its surrounding thought process lead us to a very dangerous place. Basing who we are on how we are perceived and subsequently valued by others diminishes the inherent worth of a person to that which is easily disposed of.

Listening to women is a privilege I have never taken lightly. I respect and honor that they choose to trust me with such private information. I hold that trust dearly and have never breached the confidentiality of a patient. In the pages that follow, I am going to share general scenarios under aliases to provide insight into the abortions of several of the women I've met. As I do, I will lead you not only through their experiences regarding abortion, but will share important medical perspectives, followed by God's truth on the matter. You'll also find woven into these pages personal reflections, as remembering the experiences of others brings me to share my abortion story and how it has impacted my life.

...and now, here I am, sitting in a clinic with other girls trapped in the same horrible nightmare, waiting for the door on the other side of the room to open for me. "Marissa, you can come in now." Slowly, I push myself out of the chair and walk through the open door....

My desire is that this book will not only serve to enlighten and educate but to encourage, instruct and ultimately heal. It is my hope that it may encourage those who are contemplating the abortion decision to choose life. For those who did not, it is my prayer that it will begin a journey from the brokenness of the abortion experience back to the promise of God's purpose for their lives.

PART ONE: *Reality*

Reality Of A "Procedure"

Abortion. Termination of Pregnancy, or "TOP" as it is frequently referred to in medical vernacular. Regardless of the name assigned, the topic moves me to a place of passion and compassion. Not only have I had the experience personally but as an OB/GYN, I have been exposed to a multitude of women whose lives have been ravaged by it. My vocation has given me a unique perspective on how abortion affects women. I have seen women in all stages of grief resulting from this tragedy. Fortunately, I've seen the other side as well. Privileged to be involved in many cases when abortion seemed like the only way out, women have, even at the last possible moment, made a decision to continue their pregnancy.

Our culture, including much of the Christian community, has bought into the deception of "freedom of choice" for many years. I recall viewing a film in middle school regarding embryo development—animal versus human—distinctly underlying "amazing similarities" of early gestation across the species, essentially eliminating the sanctity of life. Even at that young age, the school system was taking up the role of indoctrinating the minds of youth to this worldview. Young minds that were not truly able to discern lie from truth.

I have never seen that calculated comparison of early gestation across the species hold up in the reality of life. I can assure you that the sanctity of life is never more evident than when you see a mother mourn the loss of her baby at this early stage of pregnancy. The following is the closest scenario I have personally experienced to an abortion procedure performed in the late first trimester. Each step of this procedure was gut and heart wrenching. It added an even deeper dimension to my conviction as I saw how it affected each individual involved. At this gestational age, could it be possible for anyone to deny this as a human being?

Our only option was to take this patient to the operating room to surgically remove the fetus; I thought as I maneuvered my car through traffic on the way to the hospital that morning.

Running late, I entered the holding area under the critical glances of those who watch the schedule. I walked directly to my patient and opened her chart. God checked my rushed attitude with a heavy dose of perspective. On the stretcher before me was a broken spirit. A mother who, just the day before, received news that her newly forming baby's heart had ceased to beat.

"How are you this morning, Monique?" I asked, sensing the emptiness of the question as I probed. "Doctor, I'm sad. Very sad." She wept softly. I placed my arm around her, held her for a moment and then proceeded. "I'm so sorry for your loss. I want you to know that I understand."

...When my turn was called, I was placed in a small exam room. An IV was started, and I was given an anesthetic to make me sleep. Being a nurse at the time, I remember reading the label on the bottle of IV fluids..."Brevital" and making conversation of a medical nature with the anesthetist. I remember the stark white walls in the small room, and the paucity of instruments on the table beside the bed. As I looked into the stern face of this woman who administered my anesthesia, a woman of about 50, with dramatically applied make up...I thought how severe it appeared on her chiseled face, as if she were wearing a mask....

After answering Monique's questions, we proceeded to the OR. Even I had no idea of how I would be touched in a tender, hurting place that morning. I still wonder precisely how what followed moved each witnessing heart. Did it strike a wound within them as well? In this situation, the cervix is forced open with an instrument called a dilator. A suction tube is then placed within the uterine cavity. The suction is turned on, the amniotic sac is ruptured, and the fetus is removed. As our bodies at this stage are well formed, it was necessary for me to identify each body part as it was evacuated. Despite the knowledge that this baby was no longer alive, it was indeed the most difficult procedure I've ever performed as an obstetrician. As each portion of this little person was accounted for, my heart sank deeper and deeper. Many tears were shed during the procedure—both mine and those of others in the room. I imagine that every individual present in that operating room was taken to a deep place within their soul as that little one was removed limb by limb from its place of security. Each finger could be counted, each rib clearly seen.

Reality Of Emotions

The reality of the emotional effects of abortion is clearly demonstrated when speaking to women who are post-abortion. Not long ago I sat before a young woman who had undergone an abortion just months before. "They were excellent at telling me the physical risks. What they failed to explain was how devastating it would be psychologically."

...I remember feeling alone....and that crazy feeling that I could not believe I was in this situation. I remember being caught between professional camaraderie and embarrassment—how could I have ended up in this situation? Didn't I know better? I was urged to perform the procedure as soon as possible, due to the legal limits on gestational age at that time. I was confused....no, they are wrong on the gestational age, I thought to myself. I know when I got pregnant. I don't even remember them confirming that the baby was alive....I received no counseling. On the day I arrived for my termination, I was placed in a room full of other women who were undergoing abortions that day. There was a counselor there, who invited us (not encouraged us) to talk about how we felt about the experience. The few brief comments that followed revolved around the poor timing of the situation, as it was one week exactly before Christmas...

In preparing for a message that I delivered to a newly forming pregnancy help center, I did some research into the local abortion clinic's counseling. I dialed their number, and went through their phone prompts. I found myself listening to a telephone counseling message for women who were considering abortion. Adjectives such as "gentle" and "peaceful" were used. Very briefly mentioned was the possibility of remorse after the procedure. The message clearly conveyed that most women experienced a "sense of relief" as if a burden had been lifted. I don't think the many women I have cared for would agree. Linda wouldn't. On a routine office day, Linda was a new patient seeing me for an annual exam. She was a patient of about 50, and as we discussed her history, I inquired further about her pregnancies and children. It was then that she painfully recounted her very first pregnancy. To this day, she mourns for that son.

"Dr. Ogle, I remember it clearly. I was awake during the procedure. It hurt so much, and I wasn't sure I was doing the right thing. My doctor put my baby in a can right next to my stretcher. She then looked up at me and angrily said 'Well, you were farther along than I thought. Your baby was a boy.'"

...I could detect a tone of disgust in my doctor's voice. "We don't do those. I'll give you some information on a clinic." He left the room abruptly. I was so overcome with confusion, so numb, that I quickly pushed his response to the back of my mind...

I never have, whether in my practice or among others who have had abortions, seen a woman who was lighthearted or relieved after the procedure! On the contrary, the pain of self-inflicted shame and guilt weighs heavily on their hearts and is clearly seen on their faces. Their entire countenance has been changed.

Reality Of A "Burden Lifted"

The date was tattooed within a heart on the upper back of a young woman I examined that morning. Dani had come into the office that June morning for a pregnancy confirmation and exam. She was about six weeks along. As I examined her, I gently asked about the tattoo. "Dani, what is that date significant of?" I asked. She answered boldly. "That's the date that I aborted my first child—a date that I will regret forever."

...12-18-1982. Driving to the clinic that, ironically, was located in the town where my grandparents lived. How many times had I passed this very building, brimming with excitement at the prospect of spending time with my beloved grandparents? Never would I have suspected that this building would hold for me memories of one of the most painful and regrettable decisions of a lifetime...

As our visit together progressed, Dani shared with me how she felt coerced by the father of that baby to have an abortion. She went on to tell me that she shared with her friends the emotional pain that she has suffered and encourages them to avoid the same mistake. I encouraged Dani and offered help in any way I could, urging her to seek further counseling. Several days after our visit, my nurse brought me a chart with a phone call from Dani. She wanted us to know that she had been to an abortion clinic that morning. She had already taken the first of the two pills she was given for the "medical abortion" they were performing. She was having second thoughts and wanted to know if there was anything she could do to change her decision. At that time, the answer was clearly no. Sadly, we never heard from Dani again.

Cheri lay still on the stretcher. As I approached her side and offered an introduction, she politely acknowledged my presence and purpose. Her husband sat at the bedside anxiously, awaiting words of knowledge and reassurance. Only moments before, as I was quietly tucking my son into bed for the night, I received a page from the emergency room.

"Doctor Ogle, we have a patient here who needs you right away." The emergency room physician went on to describe the patient, who one month prior had undergone surgical abortion. She had pain for several days that was dismissed by the clinic, and then improved for several weeks, able to function normally. The pain returned abruptly, and at this point was severe.

"Cheri, I'm Doctor Ogle. The emergency room doctor called me in to see you tonight. I understand you're hurting...Can you tell me what happened?" Slowly and quietly, she explained to me the onset of her pain. The topic of the preceding abortion did not enter her explanation until I inquired.

"What happened after the abortion? Did you have any unusual complications?" Cheri then described several days of pain which eventually resolved, until this recent episode. She then looked directly into my eyes, "Could this all be from the abortion?"

"It's possible. Did they tell you that you could develop an infection?" I was grateful to hear from Cheri that they indeed had informed her of this possibility. I went on to gather information about the abortion to try to discern further if this truly was related. Cheri decided to abort her baby somewhat later than most. The abortion was performed one day short of fourteen weeks gestation. Only hours from the second trimester. My mind immediately whirled to the earlier memory of the baby of nearly the same gestational age that had died in utero, and the procedure I performed to remove it. I wondered "why had she waited so long to abort the pregnancy, putting herself at an even greater risk for complications?" yet refrained from asking. Cheri was an intelligent woman, well-spoken and beautiful. She listed her faith as Christian and had a husband who was right by her side. They had been recently married, and he had a family from his prior marriage. They had made a decision to end this pregnancy after agreeing the timing was inconvenient and then deciding they were not sure if they wanted any children.

...The lounge door opened, and he walked in. Dressed in uniform, he looked so handsome and carefree. I quietly hung my head and shared the news. He had suspected that it may be true, but he didn't want to believe it either.

"Let's check it again," he replied. "Sometimes the home tests are wrong. Jake is in the lab tonight; he'll run a blood test if we ask him to."

Oh, I thought, could it be? My heart flickered with hope. With his medical background, I trusted his knowledge and clung to his opinion over my own. He drew my blood and took it to the lab. I refused the bandage he offered as not to raise any suspicions among my fellow nurses. Later that evening he returned to the hospital floor. I made a quick and quiet exit to the lounge.

"Well, it was positive. What do you want to do?" he asked.

"Do I have a choice?" I replied, no longer whispering. I had turned the possibilities over in my head hundreds of times. Remembering that he had promised earlier in the relationship that if this were to happen he would want me to have an abortion so we could continue to court, marry and, in his exact words, 'do it the right way.' After all, he was from a deeply religious family who wouldn't accept a child out of wedlock. I knew how he felt. If I kept this baby, it would certainly be the end of the relationship, and I wanted so badly for it to continue. Maybe if I complied, we would have a chance. "I guess I'll do what we talked about." I wept and felt so alone. I couldn't even see him anymore, and honestly didn't care at that point how he felt, although it seemed to me that it didn't affect him...

Thus says the Lord of hosts: "Consider and call for the mourning women, that they may come; and send for the skillful wailing women, that they may come. Let them make haste and take up a wailing for us, that our eyes may run with tears, and our eyelids gush with water. For a voice of wailing is heard from Zion: 'How we are plundered! We are greatly ashamed, because we have forsaken the land, because we have been cast out of our dwellings. Yet hear the word of the Lord, O women, and let your ear receive the word of His mouth; teach your daughters wailing, and everyone her neighbor a lamentation. For death has come through our windows, Has entered our palaces, To kill off the children— *no longer to be* outside! *And* the young men— *no longer* on the streets! Speak, "Thus says the Lord: 'Even the carcasses of men shall fall as refuse on the open field, like cuttings after the harvester, and no one shall gather *them*.'" - Jeremiah 9:17-22

Cheri's studies were inconclusive, and although we knew she had a pelvic infection the source was unclear. Of course, a complication of the abortion was always present in my thoughts as a cause. Knowing the treatment for this could be hysterectomy, we decided to observe her carefully to try to prevent removing the uterus of this woman who had no children. She was grateful that we were attempting to avoid surgery by giving powerful antibiotics, and she improved significantly.

Several days later, however, she took a turn for the worse and was rushed to surgery. As things frequently do within the clear perspective of surgery, the cause of her pelvic infection became evident. During her surgical abortion, her uterus was punctured. This area of puncture became infected, and over the course of four weeks the infection became walled off into an abscess. The uterus, now damaged beyond repair, had to be removed to save the patient's life. Now, of course, there was no decision regarding future childbearing. That option was removed with her uterus on that very day.

Gratefully, I have also been privileged to experience "near misses" with the abortion experience. Although this situation has occurred many times, one in particular clearly stands out in my mind. Jamie, a beautiful young lady of about 19 years, came into my office with her mother one spring afternoon. Jamie was pregnant, with an abortion scheduled for the next day. As I counseled them, I discovered that Jamie had a significant medical condition. Inquiring further, her history revealed that she would be placing herself at a considerable medical risk if she were to undergo abortion the following day. After counseling on her options, along with the possibility of serious medical consequences of proceeding without consulting a specialist first, she decided to cancel her appointment at the clinic the next day. She went on to see a specialist for her disorder. Having had more time to reflect on her situation, Jamie decided to continue the pregnancy. I had the privilege of continuing to see her as a patient. Not only did she have an uneventful prenatal course, but was reconciled with the father of the baby. She delivered a healthy baby boy on January 1st of the following year. In our follow-up visit, she expressed joy and gratitude at the decision she made.

...sitting at an outside table of a coffee shop, I am watching a mom and a dad holding the hands of a sweet, beautiful girl of about 5 or 6. She gazes up at them as they speak, and they smile back at her. My mind goes to that place of wondering—was my little one a boy or a girl? I have felt for many years that she was a girl, and someday I will know. One day, I will hold her hand, we will have the same joy of speaking loving words...

Jamie's situation is not unusual, and in my experience, has been the rule more than the exception. I recall another woman that I cared for when I was a resident, Chelsea. Chelsea became pregnant after having her tubes tied.

This, unfortunately, was a tubal pregnancy, and she recovered without complications. During her workup, however, it was discovered that she had large fibroids, and would require a hysterectomy. While going through the required evaluation for her upcoming hysterectomy, we discovered that, once again, she was pregnant. After, seeing the heartbeat of her child during an ultrasound examination, she made the decision to continue the pregnancy. I clearly remember that moment in the small exam room. She turned to me and said with great conviction, "If I've been pregnant twice after my tubes are tied, that must be what God wants for me."

...I remember the moment. The feeling of dread, shuddering as the unwelcome thought overcame my mind. This was it. Finally, after waiting and wondering, I had mustered up enough courage to buy the test. Finally, I had found the strength within myself to use it. Now was the time to read it. Was I ready? Having worked the night before, I fumbled sleepily yet warily. In the dim light, I could see a faint circle. No, I thought. Positive? I panicked. Surely I've done the test wrong! I double checked the directions. Nope. Clearly positive. My heart sank. A strange feeling of despair enveloped me. The moment that I was dreading was now a reality. I sank back into my bed, sobbing. If only I could have had Chelsea's courage...

Chelsea's courage was admirable. It wasn't an easy decision for her. Her husband wanted her to abort the pregnancy and was adamant about it. She was the mother of two teenage boys, and finances were not secure. She chose, against the odds, to continue her pregnancy. She believed fully that this was God's intention for her life. She progressed through her pregnancy without complication and delivered a beautiful boy. I'm happy to report that throughout the pregnancy her husband and the rest of her family rallied around her, even becoming enthusiastic in expecting what would be another son. She named her new son JEREMIAH.

PART TWO:
The Evidence Shows

A Common Surgical Procedure

Abortion is one of the most common surgical procedures in the US. I have shared with you the emotional impact of abortion on my life and in the lives of other women. What I would like to now share are some other facts about this common surgical procedure as discovered in well designed studies done by the scientific community on abortion.

Let me preface this by saying that many of these studies have not been widely published because of their social impact. The medical community has ignored many reputable studies to protect their stance on the benign nature of this procedure. The abortion issue is a "political third rail". To be considered impartial or politically correct, doctors are often being provided with biased information regarding the impact of abortion on the lives of women.

> **"Regardless of legal status, abortion remains...an intentionally caused human death. As such, clinical and research evidence suggests it is capable of causing significant symptoms of grief, guilt, shame, and trauma."**[1]

The Truth Is Not Popular

A reputable research scientist attempted to publish results of a well-designed study that concluded, to his surprise, that abortion has a negative impact on the psychological well-being of women. This study by Dr. David Fergusson, a New Zealand pro-choice researcher who was attempting to disprove abortion's psychological effects on women, found that 42% of women who have had abortions had experienced major depression within the prior four years—almost double the rate of women who never became pregnant. The study also indicated that women who have had abortions are twice as likely to drink alcohol at dangerous levels and three times as likely to be addicted to illegal substances. The study also found that the risk of anxiety disorders was doubled.[2]

42% of women who have had abortions had experienced major depression within the last four years.

Dr. Fergusson's study was well designed, spanning the course of 25 years. As mentioned, the study was undertaken with the anticipation of validating the viewpoint that abortion did not increase mental health problems, but to confirm that these problems were pre-existing. Much to the surprise of the researching team, the exact opposite was true. When Dr. Fergusson's results were presented to New Zealand's abortion supervisory committee, which ensures that abortions in the country are conducted in accordance with legal requirements, Dr. Fergusson was discouraged from publishing the results. Fortunately, despite his political beliefs, Dr. Ferguson felt that this would be "scientific irresponsibility". Ferguson himself stated that he remains pro-choice and is not a religious person. The findings were surprising to him, however in his opinion were very robust. He went on to say that abortion is a traumatic event, involving loss and grief that may, in fact, predispose women to mental illness.

"...the debate between the pro-life and the pro-choice has, in a sense, driven the science out."

"The fact is that abortions are the most common medical procedure that young women face—by the age of 25, one in seven have had an abortion—and the research into the costs and benefits have been very weak. This is because the debate between the pro-life and the pro-choice has, in a sense, driven the science out. It verges on scandalous that a surgical procedure that is performed on over one in ten women has been so poorly researched and evaluated, given the debates about the psychological consequences of abortion."[3] Fergusson then went on to experience a great challenge in finding a journal that would publish the study, very unusual for a research team who typically has research published with the first attempt. A subsequent study by Fergusson reiterated his findings.[4] Dr. Fergusson's studies are not outliers. There are numerous studies that have shown an increase in mental health consequences after abortion ranging from anxiety to substance abuse to depression to suicide. Consistent with the finding of Dr. Ferguson's study, another far-reaching and tremendously devastating effect associated with those women who have had abortions is the use and abuse of substances. In an article written by Dr. Priscilla Coleman, a professor of Human Development

& Family Studies at Bowling Green State University in Ohio, she offers the most comprehensive assessment of mental health risks associated with an abortion that can be found anywhere in published medical literature. In brief summary, the increase in risk for anxiety disorders was 34%, depression 37%, alcohol use/abuse 110%, marijuana use/abuse 220% and suicide 155%.[5]

....and again, here I am...seeking to numb the pain. No longer is it just a habit that goes along with partying with friends...but it's now at home...every day...people are asking me, and I am asking myself...why do I want this? Why do I need it? I see myself teetering on the ledge of the abyss of alcoholism...I have to stop...

Death From Natural Causes

Many studies have also concluded that induced abortions increase natural causes of maternal death. On their website, the American Association of Pro-Life Obstetricians and Gynecologists (AAPLOG) examines a study published initially in one of this discipline's peer-reviewed journals: a 14-year study of all the women with a pregnancy event in Finland from 1987-2000. AAPLOG notes, "In this 14-year study of over 1.2 million pregnancy events, women who chose to abort their pregnancies died of ALL CAUSES combined (disease, suicide, accident, homicide) at a rate 3x higher than women who chose to deliver."[6]

Post Abortion Stress Syndrome

Despite the many studies done clearly outlining the possible mental health implications of abortion, many pro-choice voices continue to argue against its existence. Therefore, the term has not been accepted by the American Psychiatric Association or the American Psychological Association. As a physician who understands the political motives, this comes as no surprise. Consider this quote made in Psychology Today regarding the widely accepted accepted diagnosis of PTSD, however. "Nevertheless, any event that causes trauma can indeed result in PTSD, and abortion is no exception." This article goes on to list symptoms of Post-Abortion Stress Syndrome (PASS) including guilt, anxiety, numbness, depression, flashbacks and suicidal thoughts.[7]

In an article written in the Journal of Social Issues, there is proposed four basic components of PASS. (a) exposure to or participation in an abortion experience, which is perceived as the traumatic and intentional destruction of one's unborn child; (b) uncontrolled negative re-experiencing of the abortion event; (c) unsuccessful attempts to avoid or deny painful abortion recollections, resulting in reduced responsiveness; and (d) experiencing associated symptoms not present before the abortion, including guilt about surviving.[8]

...As I left through the front door of the clinic that day there were several protesters on the front sidewalk. I could feel their stares penetrating me as I avoided their eyes. The best I could muster at that moment was anger towards them. "How can they be here at a time like this? They have no idea what I'm going through. I wish they would keep their opinions to themselves." What would I say to those people now? "Thank you for being here and standing up for what you believe in"? Or would it be "Why couldn't you have been there to stop me on my way in?"...

Forgotten Victims

Approaching the psychological impacts of abortion without mentioning the impact it may have upon men who are involved in these decisions would be ignoring a significant problem. Vincent Rue, Ph. D. refers to this as:

> **"a potential mental health shockwave of personal and relational injury".[9]**

I won't explore in these pages the exploitation of men's legal rights regarding their reproductive choice beyond saying that men are completely bypassed in the legislation. No state in our union allows a husband to be informed of his wife's impending abortion. Beyond the legal issues, however, is yet another life that can be destroyed. Consider the following:

"There are some 28 studies on men's reactions to abortion that are informative. In one study, most men felt overwhelmed, with many experiencing disturbing thoughts of the abortion (Shostak & McLouth, 1984). Research evidence suggests that men are also less comfortable expressing vulnerable feelings of grief and loss, instead saying nothing or becoming hostile. Male responses to a partner's abortion include grief, guilt, depression, anxiety, feelings of repressed emotions, helplessness/voicelessness/powerlessness, post-traumatic stress, anger and relationship problems (Coyle, 2007)."[10]

....For many months I was just angry at him. For the following decades, I despised him. Now I find myself wondering if he has endured any of the emotional consequences so many men do after they make this decision... and I ask the Lord for the strength to pray for him....

We are the hollow men
We are the stuffed men
Our dried voices, when
We whisper together
Are quiet and meaningless
Remember us—if at all—not as lost
Violent souls, but only
As the hollow men
The stuffed men
- T.S. Elliott, 1925, The Hollow Men

Fathers, siblings, grandparents...the ripple effect of the decision to terminate a pregnancy affects generations, both past and future.

...I remember lying to my mother and telling her that I was at the mall that morning, was not feeling well and had decided to rest for the remainder of the day. She had no idea that I had just aborted her first grandchild....

The one who drives her to the clinic. The providers of the services. Abortion impacts so many more than just two…and the effects are felt forever, touching many lives to come…

….."My son, there's something I need to tell you"…as we sat together quietly one afternoon. "Remember all of the discussions we have had about abortion? How it is wrong in God's eyes, but how He forgives and redeems? I need you to know that you have a sibling in heaven right now…and it is only because of God's love and redemption that mom can tell you this with the hope of you understanding"…and then seeing the face of Jesus in my 13-year-old son, who grasps the reality of the cross, and forgives…

Premature Birth & Breast Cancer

Thus far, I've only outlined studies that indicate the psychological effects of abortion. In subsequent pregnancies, a history of abortion is also associated with an increased risk of premature rupture of membranes and preterm labor. There are abundant studies that demonstrate a statistically significant association between abortion and subsequent premature birth, especially extreme preterm birth. Consider the comments included in a literature review done by Dr. Martin McCaffrey, a clinical professor of Pediatrics at the UNC-Chapel Hill school of Medicine:

"Medical journals print thousands of studies annually. The challenge is to determine which studies reach clinically significant conclusions. One study, however, even if highly significant, cannot definitively establish an association as a real risk or probable cause. If a variable is a real risk, the relationship will be reproducible in other studies. The gold standard for establishing the strength of such a relationship is the Systematic Review with Meta-Analysis (SRMA). The systematic review (SR) provides an exhaustive summary of literature relevant to a research question; it uses an objective approach for the evaluation of studies on the topic with the aim of minimizing bias in those studies included in the final meta-analysis. The meta-analysis (MA) then combines results from different studies with the intent of identifying whether there is a consistent association of a factor with an outcome.

In 2009, two well-designed SRMAs were published that reviewed the world's literature on the association of abortion with preterm birth. These ultimately incorporated a total of 41 studies in their analysis and demonstrated not only an association of prematurity with one induced abortion but a dose-dependent further increase in risk for mothers with a history of two or more abortions.

> **The first study, by Swingle et al., determined that a single prior abortion increased the risk of a future VPB (very premature birth) by 64 percent.**

The second study, by Shah et al., reported that a single prior abortion increased the risk of preterm birth by 36 percent while more than one abortion increased the risk for preterm birth by 93 percent. This latter finding indisputably established that when a woman has increasing numbers of abortions, her risk for preterm birth increases further—a dose-dependent response association. Over the last two years, large national studies from Finland and Scotland provided further evidence of the abortion-prematurity association. More recently, researchers in Canada published the results of an analysis reporting that women with one abortion were 45 percent, 71 percent, and 217 percent more likely to have premature births at 32, 28, and 26 weeks. This risk was stronger for women with two or more previous abortions. Arrayed against this overwhelming evidence of the abortion and preterm birth association, there are NO SRMAs to dispute the abortion and preterm birth association."[11]

> **Other physical effects associated with induced abortion include increased risk of breast cancer, substance abuse, preterm birth and low birth weight.**

According to the Breast Cancer Prevention Institute, 58 of 74 international studies including 18 of 23 American studies suggest that there is a positive correlation between breast cancer in women who have had induced abortions (35 of which were statistically significant).[12] Studies done throughout the world are abundant, and there have been over 50 published studies between 1957 and 2013 that have shown a positive correlation between induced abortion and breast cancer. In short, the available data show induced abortion to have a positive, significant influence on breast cancer risk approximately 10 to 14 years after its procurement.

In the first ten years after an abortion is obtained and from about 15 years onward after it is obtained, induced abortion is not shown to have a positive, statistically significant influence on breast cancer risk. This "one-shot" increase in breast cancer risk seems to indicate that induced abortion is itself a carcinogenic experience and is not merely a weakening to a woman's defenses against breast cancer.[13]

There Is No Easy Option

My discussion would not be complete without touching on the topic of medical abortion. This deceptively simple option is frequently overlooked when considering the medical dangers of abortion. What most people are not aware of is the incidence of severe adverse reactions with abortion drugs. Interestingly, there is a voluntary reporting of severe adverse events to the manufacturer. This leaves it up to the provider to decide which events are significant enough to report to the drug company, who in turn determines if it is reportable to the FDA. Furthermore, the clinical follow-up after administration of this medication is at the discretion of the provider who gives it.

Considering the potential adverse events can be life-threatening or even fatal, the voluntary status of reporting is medically irresponsible. A 2006 analysis of adverse event reports (AER) released by the FDA described five deaths, 42 life-threatening hemorrhages, 46 serious or life-threatening infections and 17 undetected ectopic pregnancies (pregnancies located outside of the uterine cavity, a contraindication for medical abortion). This same group of 607 cases required 513 surgical procedures for follow up, 235 being emergent in nature with 93% of those performed to control hemorrhage. The AERs discussed in this study relate to the use of mifepristone in otherwise healthy young women and document a significant risk of severe, life-threatening, or even lethal adverse effects.[14]

A U.S. post-marketing adverse events summary through 4-30-2011 published on the FDA website states that 14 deaths have been reported.[15] Again I will emphasize that this data is limited to voluntarily reported cases. What is left to question is how many cases have not been reported. Most recently there has been an increase in medical abortion with an off-label use of a medication commonly used in the treatment of ulcers.

These abortions, when failed, have led to significant numbers of infants with severe defects of the head and face due to compromised blood flow to the developing fetus. In a culture that so highly values physical perfection, where will these babies fit in?

PART THREE:
God's Perspective

> "For you created my innermost being; you knit me together in my mother's womb. I praise you because I am fearfully and wonderfully made; your works are wonderful, I know that full well. My frame was not hidden from you when I was made in the secret place. When I was woven together in the depths of the earth, your eyes saw my unformed body. All the days of my life were written in your book before one of them came to be." - Psalm 139:13-16

God's Relationship with the Unborn

Many scriptures of the Bible indicate that God has a relationship with the unborn. I remember sharing with my son one advent season many years ago that Elizabeth's baby leaped in her womb, and Elizabeth was filled with the Holy Spirit as soon as she heard Mary's greeting, as told in the book of Luke.

> "When Elizabeth heard Mary's greeting, the baby leaped in her womb, and Elizabeth was filled with the holy spirit. In a loud voice she exclaimed: 'Blessed are you among women, and blessed is the child you will bear! But why am I so favored that the mother of the Lord should come to me? As soon as the sound of your greeting reached my ears, the baby in my womb leaped for joy!"
> - Luke 1:41-44

> "Before I formed you in the womb I knew you, before you were born I set you apart; I appointed you as a prophet to the nations."
> - Jeremiah 1:5

> "This is what the Lord says—your redeemer, who formed you in the womb: I am the Lord, who has made all things, who alone stretched out the heavens, who spread out the earth by myself."
> - Isaiah 44:24

Set us apart! Imagine that—set apart by the God who has made all things, who alone stretched out the heavens, who spread out the earth by himself! He formed us purposefully in the womb and then set us apart to accomplish specific things for Him. When He formed me in Jeanne's womb, He knew I would be called Marissa. He knew I would have two children. He knew one would join Him much too soon, and knew the other would bring me great joy and hope. He knew that I would share my story of redemption, and He knew that you would be reading this chapter. Consider his intimate knowledge of you, reader, as you contemplate what He has written.

> "The Spirit of God has made me; the breath of the Almighty gives me life." - Job 33:4

> "Your hands made me and formed me; give me understanding to learn your commands." - Psalm 119:73

On Being Forgiven

The Bible refers to sin as a crimson stain. Crimson was a dye that was formed in ancient times using the cochineal insect. What makes this analogy so powerful is that a crimson stain was impossible to remove. Many will feel the same about the sin of abortion in their lives. I know this from a very personal perspective! I can boldly answer that I have sought the full forgiveness of God through the blood of Jesus Christ who died for my sins. To receive this forgiveness is obedience. For those of you who have had the life-changing experience of abortion and have not already taken this step, I urge you to seek the full forgiveness of God that He has to offer so freely. True healing can only begin after this forgiveness is received.

> "Whoever conceals his sins does not prosper, but whoever confesses and renounces them will find mercy." - Proverbs 28:13

> "The Lord is close to the brokenhearted and saves those who are crushed in spirit." - Psalm34:18

> "Who is a God like you, who pardons sin and forgives the transgression of the remnant of his inheritance? You do not stay angry forever but delight to show mercy. You will again have compassion on us; you will tread our sins underfoot and hurl all our iniquities into the depths of the sea." - Micah 7:18, 19

> "As far as the east is from the west, so far has he removed our transgressions from us." - Psalm 103:12

I'm not claiming that this process will be easy. I am saying that any obstacle we encounter as we seek this forgiveness is one that we construct with our very own hands. For years, I turned from anything remotely associated with God. "...if He didn't like me before, I'm sure He hates me now," I would reason. "Well, no matter, I don't even believe He exists...." I couldn't believe that I could ever be forgiven for what I had done. But the truth is this:

> "For I will forgive their wickedness and will remember their sins no more." - Hebrews 8:12

You ARE forgiven, whether you feel that way or not. All the while I was measuring His capacity to love and forgive from my own tremendously limited perspective. Finally surrendering that stubborn frame of mind was the beginning of a journey of knowing God and knowing myself.

> "And I heard a loud voice from the throne saying 'Now the dwelling of God is with men, and He will live with them. They will be His people, and God Himself will be with them and be their God. He will wipe every tear from their eyes. There will be no more death or mourning or crying or pain, for the old order of things has passed away." - Revelation 21:3, 4

On Forgiving Others

07.07.07 This day will go in my journal as a day of breakthrough.

In the heat of the summer sun, in a stadium of over 60,000 people, one by one individuals representing groups of those who had participated in the pain of abortion proceeded to take the stage and ask for forgiveness:

A congressman asked women and children to forgive our nation's lawmakers for legalizing abortion.

A young man who had asked his partner to abort his child asked forgiveness on behalf of those who had requested or encouraged an abortion...

...As I walked away from the clinic, I climbed into his car. He was sleeping on the front seat. He opened the door, let me in, and we went out to lunch! The irony of that situation still astounds me. Having just terminated the life of our child, here we are sitting in a roadside diner eating

a mediocre meal and making light conversation. There was very little emotional support from him (to his credit, I suspect that he was equally as traumatized), and I was too horrified by the entire situation to share it with anyone else. He checked on me later that day by telephone. The only communication I remember after that was when he handed me a check for $105.00. The cost of the abortion at that time was $210. We divided it equally, reducing it to a financial transaction...

And I did. I forgave the father of my child...who encouraged me to proceed with the abortion so we could "get married and do it the right way".

A physician formerly employed to perform abortions by Planned Parenthood asked forgiveness for the shed blood as she cried on the stage.

...I remember being in another, larger room, in a row of stretchers with other women. There was a nurse at my bedside who was slapping my face and demanding that I wake up. As soon as I was coherent, I was ushered to another room, where I was given a prescription for antibiotics and a list of precautions, and discharged...

I went on to forgive those who participated in my abortion leading up to and on the day of the procedure. And I realized on that day, that even in my desire to minister to those who have had abortions, I had still not yet come to a place of full forgiveness in my life in this area. It was a breakthrough for me, as I realized that although I had accepted the ultimate forgiveness from my heavenly Father for the abortion I had, I had never forgiven those who had played a role somehow. I knew at that moment that this was another time of healing for me. I learned that once again the power of forgiveness brings peace to the one who forgives. My healing was being hindered by an unforgiveness I didn't even realize existed. I needed that day, 070707, to experience the power of forgiveness fully in my life. On behalf of myself and all women who had undergone an abortion, I granted forgiveness to the lawmakers, the doctors, and the fathers who contributed to the act of abortion against children and their mothers. For you, my dear reader, I pray and urge you to forgive them as well. Release them from the chains of unforgiveness, whether you know it exists or not. In doing so, you will free yourself, as I did on 070707.

"If anyone has caused grief, he has not so much grieved me as he has grieved all of you to some extent--not to put it too severely. Now instead, you ought to forgive and comfort him, so that he will not be overwhelmed by excessive sorrow. I urge you, therefore, to reaffirm your love for him. Another reason I wrote you was to see if you would stand the test and be obedient in everything. Anyone you forgive, I also forgive. And what I have forgiven—if there was anything to forgive—I have forgiven in the sight of Christ for your sake, in order that Satan might not outwit us. For we are not unaware of his schemes." - 2 Corinthians 5-11

"Get rid of all bitterness, rage and anger, brawling and slander, along with every form of malice. Be kind and compassionate to one another, forgiving each other, just as in Christ God forgave you." - Ephesians 4:31-32

"Therefore, as God's chosen people, holy and dearly loved, clothe yourselves with compassion, kindness, humility, gentleness and patience. Bear with each other and forgive whatever grievances you may have against one another. Forgive as the Lord forgave you. And over all these virtues put on love, which binds them all together in perfect unity." - Colossians 3:12-13

The Bible not only suggests but tells us that as Christ forgave us we are commanded to forgive others. In no other way can we receive the full power of our being forgiven or the peace of forgiving another. I encourage each of you, sweet readers, especially those who have been touched by abortion, to ask God to show you not only where you need to seek repentance but where you need to offer forgiveness to others as well. Then, purposefully forgive. Attach something to it that will make it a deliberate act. Get on your knees. Write it down. Ask someone you trust to stand in the place of the one you are forgiving and offer that forgiveness to them. Make it real!

On Forgiving Oneself

Why, in this case, is it for so many so much easier to forgive others than to forgive oneself? To continue to look back with hopeless regret and condemnation because we just don't feel that we can be worthy of Christ's complete forgiveness? Do we feel that we have not paid enough of a price to absolve us of our sin? Why do we fall prey to the luxury of holding on to our pride instead of forgiving the one who stares back at us every day as we gaze into the mirror? I believe this very unique and destructive face of unforgiveness is a combination of not receiving God's forgiveness and not forgiving others.

...I sit, spending time with a trusted friend with the same story of abortion. We talk about the mistakes of our past, God's forgiveness and redemption, and how He uses our testimonies for his glory. As we share, she asks that question. The one I have asked myself so many times, "Marissa, do you ever struggle with forgiving yourself? Even knowing beyond a shadow of a doubt that He has forgiven us, do you ever struggle with really surrendering it? Does it mean that we don't have enough faith?"....

Her question is all too familiar, for I have contemplated it many times. Many times questioning my ability to surrender to this truth, but always feeling a reassurance in knowing that HE who has begun a good work in me will be faithful to complete it. Indeed, this wavering has nothing at all to do with the assuredness that God's forgiveness brings but stems from the ongoing surrender of forgiving myself.

> "Behold, I am doing a new thing! Now it springs up, do you not perceive it? I am making a way in the desert and streams in the wasteland." - Isaiah 43:19

How do we fully forgive ourselves for what we have done? In a fallen world, we are conditioned to believe that everything we have comes with a price. But He is above all things, and His ways are not our ways. Nothing can be added to his sacrifice for us! So we surrender, trusting that we are forgiven, whether we feel forgiven or not, and the price has been paid. His forgiveness is full and perfect. When it comes to forgiving self, we must remember that it is a choice made out of obedience to Christ's command to forgive. Again, we may never "feel" forgiven, but our feelings are subject to our sin nature. Do not make the mistake of remaining chained to your past, as if you have not yet suffered enough for your sin. To continue to hold ourselves in a place of guilt is to deny the power of Christ. Although our hearts are broken for the little ones we let go too soon, our joy can still be complete when we step out, grasping the hand of Christ in this journey.

> "Give, and it will be given to you. A good measure, pressed down, shaken together and running over, will be poured in your lap. For the measure you use, it will be measured to you." - Luke 6:38

> "For in the same way you judge others, you will be judged, and with the measure you use, it will be measured to you."
> - Matthew 7:2

These words were spoken by Jesus to caution us to show mercy and forgiveness toward others. As I read them and know that He has taught me to forgive others fully, I then turn to that one who gazes back at me and apply the same measure to her. Full forgiveness. Seventy times seven!

"...Lord, as I have navigated through the nine-year journey of writing this book, You have torn down within me places that I had exalted against You. Pride, selfishness, condemnation of others for the same decisions that I have made myself. You have torn down self-hatred, isolation, and a stubborn self-will, all so I would return to You, your ways, and grow into the person that You have created me to be....and that I would forgive myself...."

On Grieving

...Walking one day, merely two weeks postpartum. My newborn son cradled peacefully in the stroller. My path takes me through the campus of a church near our home. I come upon a memorial to the unborn, and I find myself sitting down, weeping. Why, Lord, have you seen fit to bless me with this beautiful son after what I did? Thank you for your plan of redemption, for a new life, and for second chances...

"Godly sorrow brings repentance that leads to salvation and leaves no regret, but worldly sorrow brings death." - 2 Corinthians 7:10

Grieving is vital in the healing process, the significance of which eluded me for years. One must remember that as mothers we are created with such deep and lasting bonds with our children that we can continue to experience grief over even those who die in abortion, despite the forgiveness Christ so freely and completely gives us. It is natural to mourn for the ones we have lost, but be reassured that the condemning grief so frequently experienced after abortion is, because of Christ, replaced by a hopeful grief in acknowledging that your child is in Heaven where every tear has passed away. I know my little one is rejoicing with the Lord, anxiously awaiting my arrival. I love her tremendously, and I anticipate the day that we will embrace for the very first time, a privilege I did not have on this earth.

...Easter, 2013. I am experiencing feelings of sadness so deep, yet cannot grasp why. On a day so joyous, a day to celebrate the summation of all things real to my faith—the day that celebrates the resurrection—within my joy is sadness. And it only seems to be this time each year that I feel this yearning and loneliness for a family that isn't present with me. It isn't until several days later that I realize that I am grieving for the part of my family that does not exist on this Earth because of my decision to abort my first child. The children of that child, and theirs as well. A vacuous, gaping absence that shakes my spirit to the core when I finally identify its source....

Yet, there is hope:

"He tends his flock like a shepherd, He gathers the lambs in his arms and carries them close to his heart, He gently leads those who have young." - Isaiah 40:11

...And He gathers me in his arms and carries me close to his heart....and through the comforting words of dear friends I hear that God has a plan, and He is sovereign, and there is a purpose for every life. And that He knew that my first born would live with him in heaven, and He

is using that in the lives of many others. And God draws me even nearer still, in the words of another friend who reminds me of the future generations that will be continued because I was faithful to tell my story of depravity and redemption...

"Come, let us return to the Lord. He has torn us to pieces, but He will heal us; He has injurd us, but He will bind up our wounds."

- Hosea 6:1

On Still Healing

Those of us who have an abortion as a part of our story know that it is something that we will carry with us for all our days on this earth. But we MUST remember that perfect redemption through Jesus Christ is the truth with which we are called to measure our lives against every day. His mercies are new every morning, and every morning we need to equip ourselves with this truth: although we may find ourselves revisiting the deep waters of grief and regret, we surrender to the truth and hope of perfect reconciliation of these feelings when we graduate to life eternal with Jesus Christ! Yes, on this side of eternity we are STILL HEALING, but that healing will one day be made perfect, upon that joyful return to Zion.

"When the Lord restored the fortunes of Zion, we were like those who dreamed. Our mouths were filled with laughter, our tongues with songs of joy. Then it was said among the nations, 'The Lord has done great things for them.' The Lord has done great things for us, and we are filled with joy. Restore our fortunes, Lord, like streams in the Negev. Those who sow with tears will reap with songs of joy. Those who go out weeping, carrying seed to sow, will return with songs of joy, carrying sheaves with them." - Psalm 126

The chains are broken.
Amen.

Why I Share…

Those years following the abortion were tumultuous, as I navigated the waters of pain, regret, and hopelessness. One bad decision led to another, until God, in his mercy, beckoned me to him to fill the empty places. During that journey, I began to learn about God's mercy and grace and began to deal with the emotional and spiritual consequences of my abortion. I eventually met my husband, who became a grounding force in my life. Shortly after we were married, I was accepted into medical school. My spiritual growth continued, initially at a slow pace, but soon accelerating. As I began to mature in my faith, I could perceive God's very still and quiet voice. I also discovered that God's voice sometimes commands unwelcome requests! It was then that I felt the Lord prompting me to share my story. It struck a pang of fear in my heart. The abortion was my deepest and darkest secret. I had only shared it with two or three people. How could I share it publicly? I chose to ignore the voice and pressed on with my studies. Was I directly disobedient to God or was He just beginning to work on my will at that time? I believe He was planting a seed that would be harvested sometime later. After that encounter, the prompting to share my story returned only rarely and fleetingly, but each time it arose to the forefront of my heart and mind, I would rationalize why I was just not the right person to speak out on the topic. I had many excuses, and with each excuse the prompting would fade temporarily, but never completely. The years passed as I was caught up in completing medical school. Despite my reluctance to share what I considered to be my darkest secret, those life experiences molded my path, even contributing to my choice of specialty. God brought many of his saints to encourage me through medical school and residency, but several I remember as having a lasting impact. One professor, who, in the midst of a challenging time of my oncology rotation, shared this scripture:

> "For I know the plans I have for you," declares the Lord, "Plans to prosper you and not to harm you, plans to give you hope and a future." - Jeremiah 29:11

Those words of truth encouraged me endlessly in trying and difficult times. I began to reflect more on the plans He had for me, and trusted that despite the stresses of residency, He was in the midst of the struggles and would be my strength.

It took time, but I finally obeyed. God was asking me to step out and share the darkest times of my life with others, to help them to come out of the darkness and into the light of truth. I struggled, with tears, denial, and ultimately came to the point that I understood that the only way for others to see the fullness of God's glory in my life was to share openly about where I had been before I met Him. He wanted me to speak to women on the topic of abortion. The fullness of what God had done in my life could not be appreciated by others until I confessed openly where I was before I met him. I then knew beyond a doubt that God's purpose for me was to help women see the truth about their worth the way God sees it. What was once a dark secret in my life could become simply a reference point, as I share the truth, a truth that could restore brokenness and set many captives free.

The Bible speaks of young women of marrying age owning a jar of alabaster, a type of marble, containing a precious and very expensive fragrance. The jar was permanently sealed to protect its contents. This valuable jar was part of her dowry, and upon accepting a proposal for marriage a woman would then break the jar and pour the fragrance over the feet of her betrothed as a sign of honor. In the Gospels (John 12:3, Luke 7:36-50, Matthew 26:9-13) the story is told of Mary of Bethany, who breaks her alabaster jar to honor Jesus and prepare him for his burial. This story has often been used an analogy symbolic of pouring out our most prized possession; frequently a part of our lives that we may keep sealed and hidden, at the feet of Jesus.

I began to understand that my "secret" was, indeed, the content of my alabaster box, and if I shared it as part of my testimony, He would use it for His glory. I felt that by releasing this to Him, I would be making the ultimate sacrifice, pouring out my alabaster jar. It was something that I had kept deep in my soul for more than 20 years. He impressed upon my heart that sharing this and its effects would not only bring Him glory but would speak into the lives of women, allowing the healing to begin.

> "And we all, who with unveiled faces all reflect the Lord's glory, are being transformed into his likeness with ever-increasing glory, which comes from the Lord, who is the Spirit."
> - 2 Corinthians 3:18

Each time I share my testimony in a way that gives glory to God, it is another triumph over the enemy's plan for my life. I could have continued to allow the enemy to fill my heart and my mind with lies—lack of worth, destroyed emotions, guilt over ending another's life, and general hopelessness. Instead, I chose the freedom of accepting God's forgiveness for everything I have done. I have allowed Him to work in my life to create, from self-destruction, a story of victory, triumph, and freedom.

Dear One, if you have experienced abortion and this book has touched you, or you have not yet come to the place where you are completely forgiven by the grace and mercy of Christ, I ask you to take a few moments and pray the following prayer:

> *Lord, I'm confessing my abortion to you, seeking restoration. You know this part of my life already, but I want to surrender it to you.*
>
> *I ask for your forgiveness. I thank you for your grace and your mercy.*
>
> *I give you the hurt, the brokenness, the bondage that I've endured since that time. I accept your forgiveness and await the healing process to begin.*
>
> *I believe that the blood shed on the cross by your Son's sacrifice can give me complete and total healing. Thank you, Lord, for the sacrifice of your Son for my sins.*
>
> *Thank you, Lord, for keeping my baby with you in heaven, where your light always shines, and sadness is not known.*
>
> *In Jesus' name, Amen.*

The following pages contain letters written from several perspectives by those whose lives have been affected by abortion. Perhaps one of them will sound familiar to you. Please read on…

Hello My Precious Child,

I want you to know that I love you. How could that possibly be true if I was willing to take your life, a part of me, to avoid any consequences for my actions? I want you to know that you didn't deserve what I chose to do to you. It took me eight excruciating years to be able to acknowledge you, and it has taken nearly 13 more years of continued healing to be able to write you this letter. The truth is that my choice to rip you from my womb DID have tremendous consequences, ones I still experience to this day.

My life was completely altered the day I watched that pregnancy test turn positive. There were decisions to be made but I couldn't see past the debilitating fear and terror I felt at failing you and myself. I couldn't see any hope for life if I held you in my arms and attempted to be your mother at the age of sixteen. I didn't want to be loved and married because he "had" to. I didn't want my dreams to be extinguished. I didn't think others would accept us when we walked into any public arena. I didn't want to fail all of the extremely harsh standards I had put on myself. I didn't want a public representation of my sin because I couldn't face the shame. I also didn't believe I could physically deliver you.

The enemy whispered every lie possible into my ear. I believed that you weren't anything but tissue. He convinced me that sweeping you under the rug by destroying your earthly body would wipe the slate clean. It would be done and finished, and I could get on with my life and try again when I was older or more prepared. And so, I went completely numb. I shut off all tears and all emotions and I walked into that sterile, cold clinic and let doctors and nurses rip you from me.

What I didn't know, that side of things, was that it was only the beginning of destruction for me. I destroyed my opportunity to meet you here on earth and hold you in my arms. I denied you life and a chance at everything God had for you. I took matters of life and death into my own hands and there are always consequences for that. I did not have the humility to face my actions. I severed an entire limb of my family tree for decades and decades. And ultimately, I destroyed my ability to look myself in the eye for many, many years. I couldn't stand my own reflection.

I stopped crying after that day. I pretended nothing happened and that you were nothing to me. I believed that I could go on just as I was. None of that was true. Whoever and whatever told me I could outrun the destruction to my heart were foolish. They failed to tell me that I would continually doubt my ability to be loved from that moment on. They didn't tell me that I would not have the ability to face myself. I couldn't know that I would fight, daily, the ability to not end my own life for the atonement of yours. I didn't know that everything after that day would seem like a lie, and everything good that came wouldn't be accepted because of my own self-hatred.

I didn't speak of you to anyone. I believed the lie that if I didn't speak about that experience, it would go away. It could be stuffed away and forgotten. That was a bald-faced lie from the enemy. He comes to kill, steal and destroy, and that is exactly what he set out to do. I am responsible for every decision I made that led me to that day and since. I also know now that the enemy helped me get there. God has given me eyes to see this. God rescued me from myself before the enemy completely destroyed every ounce of me.

God helped me to see that you are at home in heaven with Him in total glory. You have forgiven me and are waiting to meet me one day. You love me. You will run to me and greet me with open arms. He loves me. I am forgiven of what seems the unforgivable. I walk in peace, light and healing, but I have to continually work at forgiving myself. Some seasons are better than others. I still know I missed every opportunity I had to see you giggle and smile and learn and grow. I missed the opportunity to receive your love and give mine to you.

I want to share our story with anyone who will listen. If we can use our pain to give life to a precious baby who would otherwise not be given that, then my pain is worth it. God gets all of the glory for redeeming my life and restoring my shattered and broken heart. I am certain that God would have seen us through each day had I not thwarted His plans for us. He would have walked beside us and gotten us through everything I thought I couldn't do on my own. For now, I have to do what I can here to be used and to hold onto my hope of Heaven and being reunited with you in glory. My heart longs for that day. I can't wait to meet you. I miss you all the time.

Until then, my love for all my days,

Your Mother

Dear Love of My Life,

My heart breaks for the mistakes we have made, for the sins we have committed. When I met you, I knew all the deep dark secrets of a young man's heart. I knew the lies and lust that had filled my life before we met. I knew that you were too good for me. You, on the other hand, felt exactly the opposite. You said so yourself. I'll never forget the tear-filled night shortly after we met that you confessed your previous pregnancy and abortion to me. I could tell that you were doing so out of self-preservation. I knew you didn't want our relationship to grow even more only to be severed when your sin was revealed. I was surprised by you. I wasn't surprised to find out that you weren't perfect. It was just that I had never met someone willing to be so vulnerable. I knew in my heart that I was and had been a sinner, and I knew that sin was sin. I knew that I had to forgive you. In addition to that, my appreciation for what you were doing led me to want to forgive you.

But it hasn't been that easy. I can't count how many times I have wondered about how this all turned out. I look back now, years later, at all the joy we have shared and the children we are raising. I should only be thankful. But at times I feel unsettled in some strange way, like I'm an intruder in a story that wasn't meant for me. I think about what might have been if you had made the right decision. And then I take the bigger picture into account. And I see the enemy at work in our lives from the beginning. I know that he wanted to trap you then, and I realize that he wants to trap us now. God is love, and the enemy is anything but. He wants us to bottle up. He wants shame to overwhelm you and fear and insecurity about my life to overwhelm me. He wants us to remain silent while he unleashes the same attack over and over and over.

I know that in my heart I have loved you since I met you. What led to that is only the grace of God working in our broken lives. I can't count how many times over I might have put a teenage girlfriend in the same position that you were in if only a few details were changed. And how many of those times would have ended up the same way? Only God knows. I know that we hurt from where you have walked. I know also that we have hurt from where I have walked. And I know that God alone can redeem all of it.

I have seen first-hand the pain that you have lived trying to find His and your forgiveness for all of it. I am so thankful to see you opening up with others about it now. I only hope that someone somewhere can heed this story and avoid this pain in their future. Thank you for being willing to speak out and bring the light into a dark place.

I love you forever,

Your Husband

To the one who holds my heart though we have never met, David Lee,

I imagine you are the most beautiful baby boy I would have ever laid my eyes on. When I see children your age, I think about how precious you would have been, and it makes me smile. Though it breaks my heart to be separated from you, you are where I long to be—in the arms of Jesus. There is no better place, Son. I wish you could have learned that on Earth like I have. There are so many things I would have taught you, David. And I would have loved you in a way no one else could have ever matched. I would have taught you how to play basketball and how to treat a woman. I would have been your biggest fan and greatest encourager. You would have probably done no wrong in my eyes. Oh, how I would have loved you, sweet boy. Forgive me for not giving you a chance. I hope you are playing with your Granddad and name's sake—I bet he's pumped to finally have a boy in the family...

If you are watching me down here, know that I will never forget you, and I am trying to use our story to help other mommies like me. But there are not many like me, David. At least it doesn't seem like it. There are not many who already know the beauty and love of Christ and still choose death for their children... I remember being called into a room with two other women while nurses explained what was going to happen after we took our pills. I think we were all pretty numb at that moment. I wanted to say something to those women. I wanted to let them know that healing would come if they just gave their lives to God—but then, if I really believed God was faithful, why would I not believe He would be faithful through my pregnancy? So I remained silent...

But I will not be silent anymore. I will not be silent about the choice that no one tells you is FOREVER. About the tears that are unceasing and you're not even sure why. About the hatred that builds up in your heart and spews out like poison to the father of the baby who promised you could start over but now wants nothing to do with you. About the memories that stain your soul and burn through your mind until you believe the only way out is death...

No. I will not be silent anymore, David. I will let these mothers know that choosing to abort you has affected every area of my life and relationship with Jesus. Even though there's healing, I am deeply scarred. And even though I am in a deeper love relationship with Jesus than ever before, the fact that I will never get to hold you or laugh with you or watch you grow up remains.

It's no secret that under our circumstances life with you would have been difficult. But life without you—always carrying in my heart the truth that I murdered you—is far more difficult than anyone could have prepared me for. And these mothers MUST know. And maybe, if you and Jesus are praying while I'm talking to them, maybe they will choose life and life more abundantly.

I love you, David Lee. Tell Jesus thank you for me. For everything. And dance with Him, David. Dance with all your might because I know that's in your Spirit because it's in my Spirit, and it's in your name.

I love you, Son.

Dear Hope,

31 years later. You would be 30 now. All of this time, all of the memories, all of the years of pain. Many years of releasing and surrendering to the Lord our story, so He could use it for good, for his glory. And only now, 31 years later, am I giving you a name, writing you a letter. Oh, I've heard it many times, how helpful it would be. But for some reason, I've never been able to do it. Was it fear? Grief? Why? Do I need to explain my decision, or do you already know?

Do you already know why I did what I did? How afraid I was, how overwhelmed, how I felt I had no options? Do you already know that I didn't love you when I made that decision, not because I chose not to, but because I didn't understand that there was a You to love? I had no idea. I believed them when they said you were nothing but a lump of tissue. But at 20 years old, would I have done anything differently anyway? I was in the midst of such a love-hate relationship with myself. I can't say.

Do you already know that I was searching for significance in a relationship with your father, just looking for a place to feel loved and secure, and felt rejected when he found out about you? My own feelings of worthlessness drove my behaviors. To react, be guarded, be transient, and abandon others. Especially you. Not knowing why. Years later, I would realize who you really were. That you were really my daughter. That we would have made it, come what may. And I miss you.

Do you already know, sweet one, that after welcoming you into his kingdom, God sought to rescue me? Beckoning me, chasing me, courting me? And when I finally realized it all, the times of struggle and tug-of-war finally led to a surrender of my life to Him.

A surrender of guilt and pain, and a deep knowing. Knowing many things about Him, about me, and even about you. Knowing that you are with Him in heaven. That you forgive me. That you even love me. That you will be in that great cloud of witnesses one day to welcome your mama when she arrives in heaven. That one day, I will hold you for the very first time.

Do you already know, my precious child, that our story has been used by the Lord in mighty ways for many years? Do you already know that it has even saved others from our pain? Of course, you must. I wonder if you and the Savior ever get to sit together and rejoice in the lives that have been healed, and the chains that have been broken. I bet you do, and I rejoice with you. Someday, because our Savior has made it so, we will rejoice together, all of us, as we see the others that He has touched with our story. And they and their children and grandchildren enter into his eternal life as well.

Until then, my precious Hope, know that you are held dear and cherished deep inside my soul. That, as your great-grandpa Artie used to say, "I love you with all of my heart".

Your Mom

Works Cited

1. Vincent M. Rue, Ph.D. , "The Hollow Men: Male Grief & Trauma Following Abortion", U.S. Conference of Catholic Bishops, 2008. http://www.usccb.org/about/pro-life-activities/respect-life-program/rlp-2008-the-hollow-men-male-grief-and-trauma-following-abortion.cfm

2. Fergusson, David M., L. John Horwood, and Elizabeth M. Ridder. "Abortion in young women and subsequent mental health." Journal of Psychology and Psychiatry 47.1 (2006): 16-24. Web. 17 Aug. 2015. <http://200.16.86.38/uca/common/grupo54/files/new_zealand_abortion_study.pdf>.

3. Hill, Ruth. "Abortion Researcher Confounded By Study." The New Zealand Herald 5 Jan. 2006. Web. 16 Aug. 2015.

4. Fergusson, D. M., L. J. Horwood, and J. M. Boden. "Does abortion reduce the mental health risks or unintended pregnancy? A re-appraisal of the evidence." Aust N Z Journal of Psychiatry 47.9 (2013): 819-27. PubMed. Web. 10 Feb. 2015. <http://www.ncbi.nim.nih.gov/pubmed/23553240>.

5. Coleman, Priscilla K. "Abortion and mental health:quantitative synthesis and analysis of research published 1995-2009." The British Journal of Psychiatry 199.3 (2011): 180-86. Web. 16 Aug. 2015. <http://bjp.rcpsych.org/content/199/3/180>.

6. "Induced abortion and natural cause maternal death." ProLifeOBGYNS. AAPLOG, n.d. Web. 16 Aug. 2015. <http://www.aaplog.org/complications-of-induced-abortion/induced-abortion-and-maternal-mortality/induced-abortion-and-natural-cause-maternal-death/>.

7. Babbel, Susanne. "Post Abortion Stress Syndrome (PASS) - Does It Exist?." Psychology Today. N.p., 25 Oct. 2010. Web. 16 Aug. 2015. <https://www.psychologytoday.com/blog/somatic-psychology/201010/post-abortion-stress-syndrome-pass-does-it-exist>.

8. Speckhard, A. C., and V. M. Rue. "Postabortion Syndrome: An Emerging Public Health Concern." Journal of Social Issues 48.3 (1992): 95-119. Wiley Online Library. Web. 16 Aug. 2015. <http://onlinelibrary.wiley.com/doi/10.1111/j.1540-4560.1992.tb00899.x/abstract>.

9. Vincent M. Rue, Ph.D. , "The Hollow Men: Male Grief & Trauma Following Abortion", U.S. Conference of Catholic Bishops, 2008. http://www.usccb.org/about/pro-life-activities/respect-life-program/rlp-2008-the-hollow-men-male-grief-and-trauma-following-abortion.cfm
10. ibid

11. McCaffery, Martin. "Abortion's Impact on Prematurity: Closing the Knowledge Gap — Clear & Concise Exposure of the Abortion-Preemie Risk." Family North Carolina (2013): 1-7. AAPLOG. Web. 17 Aug. 2015. <http://www.aaplog.org/wp-content/uploads/2013/07/McCaffreys-article.pdf>.

12. "Epidemiologic Studies: Induced Abortion and Breast Cancer Risk." Breast Cancer Prevention Institute. N.p., Nov. 2013. Web. 16 Aug. 2015. <http://bcpinstitute.org/epidemiology_studies_bcpi.htm>.

13. Lanfranchi, Angela E., and Patrick Fagan. Issues in Law and Medicine 29.1 (2014): 1-133. Print.

14. Gary, M. M. "Analysis of Severe Adverse Events Related to the Use of Mifepristone as an Abortifacient." Annals of Pharmacotherapy 40.2 (2006): 191-97. Web.

15.http://www.fda.gov/downloads/Drugs/DrugSafety/PostmarketDrugSafetyInformationforPatientsandProviders/UCM263353.pdf >.

Resources

If you have not already, I encourage you to find a safe place to be transparent, whether that is a with a friend, a group, or a professional. For women who are faced with unplanned pregnancies or are seeking post-abortion counseling, the following is a brief list of website information that may be helpful in your search:

http://www.mercyministries.org/
http://www.standupgirl.com/index.php
http://www.aaplog.org/
http://noparh.org/reading.html
http://www.nrlc.org/help/

Curious about this God who sets the captives free? Check it out!

> http://www.lifein6words.com/still-exploring/the-g-o-s-p-e-l-message-explained

CPSIA information can be obtained
at www.ICGtesting.com
Printed in the USA
LVOW05s1225040517
533239LV00026B/624/P

9 781367 241725